Ancient and Medieval Dyes

Ancient and Medieval Dyes

William F. Leggett

Coachwhip Publications
Landisville, Pennsylvania

Ancient and Medieval Dyes, by William F. Leggett
First published 1944

ISBN 1-930585-89-6
ISBN-13 978-1-930585-89-8

Cover: Medieval Patterns © Marko Beric

Coachwhipbooks.com

Contents

Introduction

The knowledge and use of color began with the dawn of civilization and is of far greater antiquity, by many thousands of years, than the utilization of wool, silk, or linen and of coarser fibers which preceded them, all of which are the successors of wild animal skins that man formed into clothing.

Color was anciently considered a spiritual necessity of equal importance to physical need of food, for all primitive people used it to protect themselves from the spells of ever-active evil spirits which they believed constantly surrounded them. This form of magic is older than a belief in gods, and, at a very early stage, magical symbols were added to former mere daubs of dye, and here we have the ancient origin of art.

The earliest dyes were quite possibly discovered by accident, and may have been stains from available berries, fruits, and nuts used as food and later from blossoms, leaves, stems and roots of shrubs, or bark and twigs of trees, and the dyeing art seems to have independently developed and been practised among primitive people of almost every country. Primitive man and his age-old succession of descendants almost exclusively used dyestuffs of vegetable origin, all easily obtained in the immediate vicinity of use, the prospective dyer merely collecting known dye plants or roots in nearby fields or forests and boiling them in hot water. They wrought within a narrow range of color confined to red, blue, yellow, green, brown, and black, with limited varients in shades and tones. Chance, aided by keen observation, undoubtedly led to the discovery of other sources of dyestuffs, so, little by little, dye knowledge increased, and, so far as it can be ascertained from a score or more of ancient writers, nearly one thousand different plants, vines, shrubs, and trees were, at one time or another, used for extracting dyes. Trade in dyestuffs began as soon as the sources of one district were recognized as superior to those used in another district, and, ultimately, this led to the elimination of many of the anciently used dyestuffs, so that of the many hundreds of original primitive dyes only a few survived to ancient and medieval times. The most important of these, divided into vegetable, animal, and mineral groups, are discussed in this book.

The Vegetable Dyes

Madder

Madder plants are of about 3.5 species, belong to the order Rubiaceae, and, in ancient and medieval times, were cultivated over an enormous area of the Near East and later in Europe. They are a hardy growth, varying in height from three to ten feet, depending upon climate, irrigation, care in culture, age of plant, and calcium carbonate content of the soil. These plants are indigenous to many countries in the tropical and temperate zones, but are more abundantly found in Asia and Europe. Roots, having the greatest dyeing value, are taken from *Rubia tinctorum* (*foliis senis*) or "dyers' root," by far the commonest of rubiaceae, and *Rubia peregrina* (*foliis quaternis*), both of which grow in the Near East, the Caucasus and in Europe.

The madder plant has a square and jointed but weak stem and, at each joint, four to six leaves grow, pointed at both ends, and about three inches long by one inch wide at the middle. The upper side of the leaf is smooth, but the center nerve, on the under side, is armed with small rough prickles, more of which grow on the stem. The branches, which usually bear flowers in June, proceed from these joints and then divide into four yellowish leaves. The fruit, which contains a round seed, is a variety of berry which, as opening time arrives, is at first a brownish shade, but soon turns black.

Each madder plant root is surrounded by many small fibers, has a yellowish-red pith, and is covered with black bark or rind. Old roots were anciently considered to be richer in pigments than younger roots, but in Europe, each plant was left in the soil for only 24 to 30 months.

The bulk of pigment is contained in the red mass, between the outer skin and the woody heart of the root, the dye being present in the form of glucosides which are quite easily separated, the most important being rubcrythric acid, which is broken down into alizarin, chlorogenin, purpurin, and sugar. Madder plant roots are normally of considerable length, but in thickness rarely exceed that of a common lead pencil, although, some varieties attain the thickness of a finger, and each root greedily pushes itself as far as possible into the soil. Although European medieval madder growers did not always consider it advisable to increase the size of these roots, often taking measures to keep them as small as possible, yet, in the Near East, it was a custom to train madder plants on frames as this procedure was incorrectly thought to increase the volume of dyestuff in the roots. European madder users endeavored to check this Levantine custom, as it also increased the waste of each root—madder roots were bought by weight—but, failing to do so, at last introduced scientific madder cultivation into Europe, and were finally able to supply a rapidly diminishing local market for Oriental madder roots.

After each root has been dug up, it was thoroughly washed in pure water, allowed to dry, either naturally in the sun, or artificially in kilns, and finely ground to powder and stored in bags or casks. In certain Near Eastern madder growing districts, ancient and medieval madder cultivators, before grinding, stored the roots for several months in underground pits as it was thought that this increased tinctorial value.

The name for madder, in the various languages, and its connotation, red, clearly indicate the color content of the plant. It appears in Arabic as *al zan*, in Greek as *erythrodanon*, in Roman as *rubia* and in German as *rote*, but notwithstanding this etymological concord, no other dye producing medium underwent such critical examination by scientists of Asia and Europe, for over 300 years, as did the madder plant, and in many of the older manuals of dyeing, much space is given to these researches.

Madder has been known from an antiquity so remote that it is not possible to determine with any certainty just where and when it originated nor are many details of its earliest culture known to us. Undoubtedly, it was first used in India, but it appears to have been equally well known to ancient Persians and Egyptians, and, considerably later, to the Greeks and Romans. Although very meagre accurate data, concerning the early history of this plant, have been handed down to us, yet cloth, very plainly madder-dyed, has been found on Egyptian mummies, in tombs of a predynastic era. Ancient Hebrew laws permitted the culture of madder solely for household consumption and strictly prohibited its growth for commercial purposes. In the first dye trade document, written in a European language, Greek, there is record of trade in this root between India and Asia Minor. It is in the form of a "periplus" or narrative of a coastwise voyage along the Red Sea in the first century of our era.

About 450 B.C., Herodotus tells us that in his day, "rubia was used to brighten the cloaks of Libyan women." If critical comparison is made of what is known today about the madder plant with what Dioscorides, that Greek medical genius and author of "De Materia Medica," a scientific work that was popular for several centuries, wrote about a plant he called erythrodanon, it will appear that he referred to madder. He tells of the "long square stem armed with a great many hooks, the leaves standing around the joints in the form of a star," adding that "at first the fruit is green, then red, and lastly black" and "the long roots, which are red, serve for dyeing cloth." He also related that this plant "was cultivated with much profit at Ravenna and Caria, where it is planted among olive trees, or in fields prepared for olive trees." Dioscorides not only knew botany but he also knew the Roman Empire, for, although a Greek, he was a medical officer who accompanied the Roman army on many of its campaigns.

Pliny, the Elder, author of "Naturalis Historia" tells of madder cultivation in the vicinity if Rome, in 50 A.D., or about thirty years before he lost his life in that famous eruption of Vesuvius in 79 A.D.

Other ancient writers, among whom was Theophrastus, the successor of Aristotle, agreed with Pliny and Dioscorides, and they even went farther when they said that erythrodanon, in the Roman mother tongue, was called rubia, and that "its red roots are used to dye wool and leather red." In this connection, it may be interesting to know that the directions which preface the several prayers and offices in the missals, called the rubric, received this name because of the practise, among monks of the Middle Ages, of writing them in red ink. With the passage of time, this origin of the word was forgotten, and is now used to indicate the directions themselves.

The red dye, which, with characteristic forethought, Alexander the Great purchased for the use of his invading army in 330 B.C. may have been derived from kermes, which was the insect found on Asiatic oak trees, and not from madder, although both forms of red dye were known to him and to his Persian adversaries. But it may be of interest to relate how red dye helped Alexander defeat the Persians because of what may have been the first important use of camouflage in war. When informed that he was faced by a Persian army much larger than his own, the youthful monarch remarked in calm contempt: "The wolves never concern themselves over how many the sheep are," so, with true strategic genius, Alexander, one night, caused the clothing of a large number of his soldiers to be dyed red, at a different spot on each garment. Next morning, when the Greek forces advanced—shall we say simulated a stagger?—to meet their enemies, the Persian leaders thought that the soldiers of Alexander's army had been pretty well damaged during the fighting of the previous day, with little opportunity for medical attention, so they may have been unduly careless when making an attack on what looked to them to be a helplessly wounded antagonist. Alexander (the Great) won that battle!

With the collapse of the Roman Empire in the fourth century A.D. and during the Dark Ages in Europe, which lasted almost eight centuries, except for isolated instances, all record of European madder ceases, for that catastrophe temporarily shut off the flow of madder and, of course, other products from the East to Europe. During the following long period of unrest, even local cultivation of madder declined, so much so, in fact, that the dye trade of the world shifted back to the Orient, where it grew into such an important element of Asiatic commerce that, for a long period, Bagdad, then a magnificent city, rivaling Byzantium, was the most important center of dye trade. For the first time in its long history in the Orient, the madder plant was actually cultivated instead of being permitted a "Topsy-like" development. This was because Eastern dyestuff dealers realized that they had now been called upon to replace a more scientifically cultured plant, no longer available to European cloth merchants, and, strange to relate, they broke free from normal lethargy and became so expert in madder culture that Bagdad even exported madder roots to India, a land which for centuries had practised a certain degree of madder plant culture.

Madder is not again referred to, in European records, until the early years of the seventh century, when it is recorded that madder "brought from the East" was cultivated at St. Denis, near Paris, which shows that at least some effort was made to revive a dye industry which for three hundred years had all but disappeared.

Soon medieval rulers became interested in madder, and in the eighth century, Charlemagne decreed that madder plants be cultivated "in these estates," which may have been his rather extensive empire, including all France, Belgium, Holland, most of Germany, Austria, parts of Italy and a section of Northern Spain, and undoubtedly, it was because of this edict that medieval peasant-farmers made a practise of growing madder in fields left fallow, because of a startling modern knowledge of the benefits which follow rotation of crops. In France this crop rotation system was afterwards officially enforced for a long time.

Grown in Holland as early as the 10th century, madder appears more frequently in the historical annals of Europe, especially in France, where it was reintroduced at the time of the Crusades, and in Italy, where a revived dye industry grew rapidly because of a demand upon it by cloth makers of Palermo, Genoa, Lucca, Venice and Florence. The noted Florentine family, della Robbia, famed as sculptors, took their name from the madder plant, which was known to them as rubia. One of their ancestors had been a Florentine cloth merchant. During the Middle Ages, brilliant red dye was, next to purple, the most favored dyestuff, for that period of transition loved bright colors and the most colorful of all cloth was the dress of men.

Having the benefit of both skill and material from the East, the Moors, as early as 900 A.D., revived cultivation of madder in Spain, and finally developed a satisfactory export trade with both Portugal and England. France, it should be recalled, had introduced madder culture about two hundred years prior to this time, but meagre available records indicate that it supplied merely a local market for dyes, and it does not appear that France either imported or exported madder roots until several centuries later.

The discovery of an all-water route to India in 1498, by Vasco da Gama, which overcame the Turkish monopoly of the Mediterranean Sea, also contributed to improve the European dye situation, and soon imports on a large scale, involving additional sources of dyestuffs, became possible. The new sea route from the Orient to the west coast of Europe reduced costs which formerly had been almost prohibitive because of so many intermediate handlings, each involving labor and commission fees, and also meant the almost certainty of arrival of goods at fairly predictable, scheduled dates. Soon after the da Gama discovery, Spanish forces, in conquered Mexico, became acquainted with cochineal, which, in that country, had for centuries been used as a red dye. Although cochineal has no place in the history of madder yet the combination of Portuguese ships, bringing dyes from the Orient, and Spanish ships, bringing dyes from America, meant that the center of the dye trade, so far as Europe was concerned, had moved from Bagdad to Spain and Portugal, for all important dyes that could not be raised, now arrived by water from both Western and Eastern sources, and were distributed from ports much nearer to places where the dyes were used.

About the year 1494, detailed instructions, concerning cultivation of madder plants, had been published in Holland and they reveal the high standards which that sturdy country endeavored to attain, although, politically, still under the heel of a conqueror. For the next three hundred years, the Dutch were the most advanced madder growers in the world. There may have been some contributing natural advantages such as the moist climate of marshy coastal districts, reclaimed from the sea and the alluvial soil of the many deltas, each of which surely contributed to make Holland ideally suited for madder cultivation. However, no practical student of economics will overlook the fact that, for at least two hundred years prior to 1500 A.D., Holland was also one of the foremost cloth-fabricating countries of Europe, and, consequently, a large importer of both wool and silk. Thus it may be inferred that it was not solely the unsatisfactory quality of Asiatic dyestuffs that led to local culture of madder. The thrifty Dutch may have had one eye on valuable exchange at Amsterdam and Rotterdam, to pay for raw silk and wool which they could not produce, but which must have been kept flowing to their many cloth factories. Dutch farmers, too, undoubtedly applauded the decision to raise and improve local madder plants, as it meant income for home folks rather than Oriental strangers. Likewise, cloth merchants, who purchased the product of the factories, rested easier, since a necessary and popular pigment that formerly faced delayed and dangerous shipment in a combination of caravan and sea routes now afforded promise of a constant supply and timely arrival.

Although not until the beginning of the 18th century did France become a serious competitor of Holland in madder sales to Europe, yet that was the only country that offered any substantial check to the long-held madder market position of the Dutch. The genesis of this French competition was solely due to necessary economic steps of preventing large sums of French exchange being sent out of the country, not only to Holland, but also to Italy and the Near East. The incentive for wide culture of madder in Normandy came from an Armenian, who had been trained in madder growing in Persia. He claimed that "a favorable conjunction of climate and soil" made Normandy "an ideal locality to develop madder."

Louis XV directed his Ministers of State to "arouse interest in madder growing among the peasants of Southern France." They ignored the primary Dutch theory of planting roots instead of sowing seeds and the seeds, which they imported from Cyprus and Smyrna, were described as being "specially treated" (irradiated?).

The coming of the French Revolution and the long period of political and economic unrest, which followed, played havoc with this thriving industry, and not until 1815 were any efforts made to revive it, and even then, it but partially recovered its former vigor. In an effort to succor a once prosperous home industry, King Louis Philippe, in 1840, ordered that the trousers and caps of the entire French army be "dyed red with madder." In so doing, he was but following the precedent of an earlier English king, Henry II, who ordered red coats to be worn by all who engaged in fox hunting, as that monarch had made it a royal sport, enjoining all who took part to wear the royal livery. As the army was in royal service, he ordered all army uniforms to be "dyed red with madder," and American school children still know these soldiers as "red coats." These moves, themselves, did not consume any great quantity of madder roots, but both the emotional French and the more phlegmatic English expressed a certain patriotism by demanding this brilliant red color in many forms of cloth both for the person and for the home. This made the efforts of both madder-reviving kings successful.

In an attempt to foster a new industry, King Charles I of England, about 1615, decreed special privileges to those of his subjects who would cultivate the madder plant, but imports from Holland and France were always required to maintain a steady supply of this red dye for English cloth dyeing. A large part of imported and local madder, used in England, passed through Norwich, making this town the most important madder distribution point in that country. Memory of this activity is still preserved in an age-old, local thoroughfare called Madder Street. The cultivation of madder in England was never commercially successful, although it had long been grown on a limited scale, and small cloth dyers purchased, what supplies they required, from local farmers who raised madder mainly to sell to herb dealers and local physicians, for many of the latter still credited this plant with some of the medicinal virtues ascribed to it by the ancients. For some reason, which may have had historical basis, England preferred green or brown as a color relief in dress, and did not make use of red to the same extent as it was used on the Continent.

Non-European countries, in which madder was grown, included Persia, Mesopotamia, Egypt, Tunisia, Morocco and India. The alizari (madder) of Arab merchants was, at one time, far superior, in quality, to European varieties of rubiaceae, and was allowed to remain in the ground for six years, for, unlike the peasant growers of Europe, who, because of taxes and higher living costs, could ill afford to permit their land to bear but one crop for such an extended time, and, therefore, introduced scientific cultivation, the Arabs depended entirely upon age for size of root and, facing less economic pressure, could afford to adjourn to their tents and wait for it.

Madder, grown in the Orient, had one favorable characteristic which cannot be disputed and this was the advantage of pure open air drying of roots, which considerably increased the value of Oriental madder when in competition with roots which had been dried in the always moist and often impure air of the usual European drying room of that period. Edward Bancroft, a celebrated English authority on dyestuffs, once slyly remarked that the difference between European dried madder roots and those dried in the Orient, was as great as the difference between a European and an Angora goat, but despite this distinguished endorsement, madder from the Near East, especially after the 12th century, never achieved marked success in Europe.

For many decades, during and after the fifteenth century, the bulk of European powdered madder was packed for transportation and sale in casks or linen sacks. In those remote days, no modern "guaranteed sample of package content" system had been devised to protect or, at least, guide the prospective buyer. Therefore, it was practically impossible for him to accurately know the exact quality of the powdered dyestuff offered for sale, as only actual use could develop that point. Thus he was entirely dependent upon the integrity of the seller. Be it recorded with pride that the majority of European madder growers were reputable and did their utmost to prevent falsification. Holland and France, as the most important madder growing countries of that era, enjoyed excellent reputations in this respect. In Holland, quality standardization was maintained by strict regulation which governed both manufacture and sale. Some of the regulations, which had been put into effect in 1537, by Charles V, were faithfully followed by the Dutch until the gradual decline of madder production, in the nineteenth century.

When synthetic dyes, in volume and variety, were finally placed on the market, madder cultivation declined all over the world and, within a decade, had entirely retired in favor of the synthetic "alizarin," for the latter was one-fourth as costly and more effective.

Today madder is cultivated, in rather limited quantities by a very few countries, solely to supply a demand of artists for a certain natural pigment of superior quality and, possibly, for some physicians who still "credit this plant with some of the medicinal virtues, ascribed to it by the ancients."

Indigo

Indigo, long regarded as the most important of all dyestuffs, is a vegetable blue dye of great natural fastness to both light and water and, as is proven by some very ancient records, written in Sanskrit, describing various methods of preparation, has been known to the people of Asia both as a dye and as a cosmetic, for over 4000 years. It derives its medieval name from *indicum*, a Latin word originally used to define all imports from India and later specifically applied to a beautiful blue dye of India, replacing the anciently used Arabic word *al-nil* which meant blue, and which is the ancestor of the modern word *aniline*.

Indigo, one of the most important and popular of a wide range of ancient and medieval dyes, remaining so until comparatively recent times, was obtained from the leaf of *Indigofera tinctoria*, a member of the order leguminosae, widely distributed in Asia, Africa, the East Indies, the Philippines and America. In very early times, natives in all countries, where the indigo plant grew, seemed to have known how to utilize its water-soaked leaves to obtain a brilliant, blue dye. Garments, found in Egyptian tombs, and others, unearthed from Peruvian Inca graves, as well as many remnants of miscellaneous primitive cloth, all testify to wide but unconnected knowledge of the indigo plant.

For many centuries, and until the discovery of America, in Europe and the Near East, all natural indigo, used for dyeing cloth and as pigment in paint and cosmetics, came from India. Periplus, in 80 A.D., recorded that indigo had long been an article of export from India to Egypt. At first, it was transported to Western Asia and Eastern Europe by long, expensive and hazardous caravan routes to the favored distribution points of Greek and Persian merchants, on the Red Sea or Persian Gulf, and later by the all-water route from India to Portugal.

Indigo dye is contained solely in the leaf of the indigo plant. Unlike other vegetable dye sources, the stalks, pods and twigs of this plant do not contain enough color to warrant processing. The actual amount of pigment obtained from each leaf is small, and this had the effect of making indigo dye one of the most expensive in medieval dyeing. Indigo is in the form of a colorless glucoside known to dye masters as indican and is soluble in water.

The shrub-like indigo plant attains a height of from three to five feet, has long slender pods and narrow leaves, and can be harvested twice a year. When ready for its simple processing, it is cut down early in the morning and, as quickly as possible, placed in vats and steeped in water for a period varying from nine to fourteen hours. It was quite early noticed that the percentage of indicum changes with the freshness and moisture content of each leaf. The liquid, which varies in shade from yellow-orange to olive-green, is drawn off to heating vats, where it is exposed to oxidation by air, by means of striking the surface of the slime-like substance with strong bamboo sticks, a process which until comparatively recent days has been little modified in India, since ancient times. This operation causes a gradual change in the color of the liquid, it becoming dark green, and finally blue. When the precipitate readily settles, beating is discontinued, and the mixture is permitted to rest for about two hours. The top water is then drawn off, and the indigo sludge led into a reservoir from which it is transferred into a large cauldron where it is heated by fire to prevent further fermentation. When cool, this sludge is filtered through strong linen or cotton cloths stretched over vessels of stone, where it remains until it reaches the consistency of a stiff paste. The paste is then formed into bars, which, in turn, are cut into the indigo cakes of commerce.

Indigo had long been used for dyeing and painting in Europe at the time of Pliny and Dioscorides, although, it must be understood that in those early days, every kind of blue pigment, separated from plants, by fermentation and converted into a friable substance, by desiccation, was ignorantly called indigo, for merchants and dyers considered as real indigo dye, blue substances, prepared in many different countries from a great number of plants, some of which were not even varieties of the same species. The Greeks and Romans appear to have powdered it for use as a paint or a cosmetic and the use of indigo for painting remained general until the invention of painting in oil. When mixed with a slate-like earth or with white earth, it could also be used as a blue crayon, and both Greek and Roman ladies used it to ornament themselves.

The real characteristics of indigo were not widely known until in the 14th century, Holland imported this dyestuff directly from India; yet even long after that period, indigo was often thought to be a mineral. In fact, as late as 1705, a British patent was granted for "obtaining indigo from mines," for it was, even then, considered, by some, to be "a blue stone brought out from India and used for painting and dyeing." However, what Dioscorides called "indicum" and what, with slight variation, both Pliny and Vitruvius called "indicum," were undoubtedly what we today know as indigo.

Both Pliny and Dioscorides, each an experienced botanist, speak of two kinds of indigo, one of which "adheres to reeds in the form of slime or scum, thrown up by the sea," and the other, Dioscorides specifically mentions, was scraped from the sides of the dye pans in the form of a purple colored scum and was collected, in this manner, in all establishments using a purple dye. Anciently it was classed as an astringent medicine, as "it cleaned wounds and was used for ulcers and inflammations." Dioscorides highly praises the medicinal properties of indigo, and some experiments, made as late as the 17th century, seem to confirm his estimate of the value of indigo "for internal use."

Venetians appear to have been the first medieval people to use indigo, for they very early realized the value of this dyestuff, both from an artistic and a commercial point of view. There is a record, dated 1194 A.D., which concerns the importation of both indigo and brazilwood from India. Although indigo was used in Venice at this period, its use does not seem to have extended to the rest of Europe, even when the decadence of Venetian industry led to an increased knowledge of their dyeing art in Western Europe and, subsequently, reached a high degree of perfection in France and England. Because Europe of that period was more favorable to woad, indigo was not generally introduced until the 15th century.

The Venetian traveler and merchant, Marco Polo, who had spent twenty-six years in Asia, at the end of the 13th century, and whose lively account of his travels is one of the most important sources of information about that era, refers to both indigo and brazilwood as something familiar to Europe. He relates that "it is won from a species of herb which is plucked out by the roots, and put into tubs of water, where it is left to rot. Then the sap is pressed out which, when exposed to the sun, evaporates, leaving a kind of paste behind it, which is cut up into small pieces of the shape also familiar to us."

In the middle of the 14th century, some Italian writers called indigo "indaco disi baldacca," suggesting Bagdad, from where it was sent to Europe, packed in hides, leather bags and sacks. It is the opinion of some historians that Hebrew dye masters, who were familiar with dyeing practices at Bagdad, first introduced indigo dye-stuff into Italy, but it is known with certainty, that the Italians were the first people of Europe to bring its use to a greater degree of perfection than did any other European people. Since indigo replaced woad as a source for blue dye, Italian cloth dyers were able to secure such beautiful blue colors that a knowledge of indigo soon spread to other countries.

However, its use as a dye, in Northern and Western Europe, was limited until after an all-water route to India was discovered, in 1498, although no large quantities were imported by water until about 1602. Even then, the influential growers and distributors of its European rival, woad, offered such opposition, both on the Continent and in England, that various laws prohibiting its use were enacted and enforced. Although there is evidence of the introduction of indigo into England about 1587 A.D., yet nearly a century and a half elapsed before indigo was imported in any large quantity. In an effort to promote the sale of indigo from his Indian colonies, the King of Portugal, in 1643 A.D., sent an emissary to England, and ordered him to demonstrate "the value of the dye called angel (anil) commonly called, in English, blue ynde, which cometh out of the Yndies, and by repute is made of the flower and first croppe, and cut on all growing there, whereof woade is made, not before this tyme practised upon woll and clothe of England." But it was called "devil's dye" and reported to be injurious to cloth fabrics.

Barbosa, who afterwards accompanied Ferdinand Magellan on his famous voyage around the world, perishing with his leader on the island of Cebu, in the Philippine group, collected much information about indigo, and also reported the prices which were demanded for it at Calcutta, in 1516.

In 1563, an Italian traveler, in listing merchandise purchased from the Portuguese by Antwerp traders, specifically mentions "indigo from the East Indies." About 1602, a Netherlands trading company was organized "to supply products indispensable to Europeans such as cotton, tea, sago and other things plenty of which they could always hope to find in India." One of these products was, of course, indigo. German cloth merchants complained that their native woad "was being banished by indigo" and traced this situation to Netherlands importers, for this dyestuff had been introduced by them into Germany about 1604. One amusing angle to this German complaint was the charge that "gold was given to the Dutch for a worthless dye, whereas our woad industry was allowed to decline." These 17th century German dyestuff merchants correctly feared that importation of indigo would ruin their profitable woad business.

Soon after the discovery of America, the profit in indigo was an incentive to cultivate this plant in the New World, for Spain was a commercial rival of Portugal, and Spanish adventurers had noted that the American Indians tinged their bodies and faces a blue-violet color and the pigments for this came from a native plant, closely resembling the indigo plant of Asia. It was mentioned by Francisco Colon (Columbus) in his "Life" of his renowned father as among the valuable products of Hispaniola, the present Dominican Republic. Further mention of this dye-producing plant, by other Spanish explorers and writers, especially those who were familiar with Mexico, seems to sustain the belief of many botanists that the indigo plants of the new and old worlds, and apparently found on three-quarters of the earth's surface, are often specifically different, yet belong to the same family. The first American indigo brought to Europe was obtained in Guatemala, but not many years passed before Mexico provided a large quantity.

With indigo entering Europe from both the East and the West, promoted by rival commercial nations of Spain and Portugal, because of the superiority of its dyeing quality, and the added fact that indigo contains ten times more pigment than does woad, it very quickly threatened to supersede woad in the dyehouses of Europe. At first, a small quantity of indigo was added to woad in order to improve quality, but soon the proportion of indigo became so large that woad was used merely to revive fermentation of indigo, and woad itself was incapable of contributing any additional color. Contemporary chemists observed that cloth could be dyed more cheaply when indigo was combined with a small quantity of woad than when the latter was exclusively used, as had been previous practice, and this led to economic repercussions, for woad farmers of Germany, France, and Italy lost a market which had been profitable for many centuries.

In 1577, an English local governing body "prohibited under the severest penalties the newly invented, harmful, balefully devouring, pernicious, deceitful eating and corrosive dye known as 'the devil's dye,' for which vitriol and other eating substances are used instead of Woad." This prohibition was renewed in 1594 and again in 1603, when the statement was added that "in consequence of the weight of this dye a pound of undyed silk, for sewing or embroidery, would produce two or three pounds of dyed silk." The reference seems to be to certain astringent juices, which, about this time, began to be used in dyeing and because of unfamiliarity with the chemical action of these new agents, when cloth was boiled in coloring vats too long, it became exceedingly tender. But even if endowed with a firm determination to discharge their civic duty, the action of these officials seriously interfered with a "trial and error" advance in dyeing art. In many countries, there was a sincere belief that indigo was actually harmful and destroyed fabrics on which it was used, a condition possibly due to inexpert use of the dye bath. As late as 1664, and in some countries even later, decrees were proclaimed against the use of indigo. But despite all such measure, indigo became the most popular dye in Europe, and as it became common, the market for woad was greatly reduced so, to restore woad sales, Germany enacted many laws prohibiting the use of indigo. A series of such laws was issued in Saxony in 1650, and in an effort to justify this action, the government included indigo as an "eating substance like 'devil's dye.'" In 1652, this law was expanded to banish indigo from all parts of the State, and in addition, it gave some exclusive privileges (not named) "to those who dyed with woad." In 1654, these Saxon laws were made imperial laws and everything ordered with regard to "devil's dye" was repeated, with the addition that "great care should be taken to prevent private instruction in the use of indigo, by which trade in woad would be lessened, dyed articles injured, and money carried out of the country."

In Nuremberg, where woad was cultivated at that time, the people went even further and passed a local law requiring their dyers to make an oath, annually, that they would not use indigo.

The steps, taken in Germany to protect the woad industry, were taken even earlier in France, where, in 1598, because of urgent representations of hardship, the use of indigo was positively forbidden. This prohibition was subsequently repeated many times until 1669, when Colbert, the French Comptroller-General of Finances, in classifying French industry for tax purposes, separated the "fine from the common dyers and ruled that indigo could be used without woad." This edict prevailed until 1737, when it was ruled that French cloth dyers "were at liberty to use indigo," and it is from this time that the valuable dyeing properties of indigo appear to have been recognized. During that era, physicians, in many countries, sometimes used indigo as a drug, but in England, Elizabethan laws denounced it as "a dangerous drug" and forbade its use, passing an act which permitted searches to burn it "in every house where it can be found." This Act remained in force for almost a century, or until the reign of Charles II. Dyeing practice in those days called for a preliminary blue dye bath for any fabric to be dyed black, and hats of that period were not considered to be properly dyed unless traces of a blue tint could be identified when the completed fabric was cut for shaping.

In the early years of British occupation of India, indigo formed an important item in the trading activity of the East India Company, but because of the crude method of native processing, this trade declined in favor of the more skillfully prepared product of America and the West Indies. It is interesting to learn that in 1747, subsidies for cultivating indigo were offered to farmers of the Carolinas, and that, a little later, the indigo crop was a very important agricultural item in Georgia, North and South Carolina.

Indigo was long used in large quantities to dye the uniforms of American and English sailors. In conclusion, it may be interesting to learn how blue was adopted as the almost universal color for naval uniforms. Other navies had official uniforms long before the British and, in 1745, some British naval officers, meeting at Will's Coffee House in London, decided that they would petition the Admiralty for an uniform in order to standardize it, as had other European navies of the day. These officers were directed to appear in person at the Admiralty, each clothed in a uniform which was designed and colored in with his individual ideas. Some wore gray with red facings, but one naval captain wore a blue uniform with white facings, and the latter was chosen by the Admiralty. The story goes that since King George II must make the final decision, the enterprising captain tactfully selected colors used in a special riding habit of the Duchess of Bedford, a favorite of the King and the wife of the First Lord of the Admiralty, who frequently rode in Rotten Row, that still fashionable ride in Hyde Park, London, named "route un Roi," or King's Road. These colors sentimentally gained His Majesty's approval.

WOAD

The woad plant, known to botanists as *Isatis tinctoria*, is classified in a group which derives its name from the Greek word *isos*, meaning equal, because it was once thought that this family of plants possessed the virtue of removing skin roughness by applying its leaves to the affected parts. The word "tinctoria" was added by Linnaeus, in 1753, to describe its dyeing qualities. It is both a biennial and a perennial plant and belongs to the family cruciferae. The stem of the woad plant is from two to five feet high, and the leaves, from which the blue dye is obtained, are orate or lanceolate and stalked. It is thought to have originated in Southern Europe, from whence it spread as far north as England and Sweden, taking on the characteristics of a plant native to the temperate zone, forming part of the mesothermic flora.

To prepare the blue dye, the newly gathered, selected leaves, cut off at their base, were quickly crushed or ground to pulp. It was quite early discovered that young leaves supplied a light blue dye, mature leaves, a darker blue, whereas fully ripe leaves provided a bluish-black pigment. After crushing, the pulp was placed in small heaps to drain, until they became sufficiently dry to cohere. Then, by means of hand kneading, they were converted into balls, each about three to five inches in diameter and weighing about five pounds. These balls were then spread on wicker trays and dried for about four weeks, in well ventilated sheds, prior to storage in a dry, airy place, pending gathering, and processing of the entire crop. After this, the complete lot was fermented. For this latter purpose, each ball was ground into a fine powder and the entire mass spread to a depth of two or three inches on the floor of an open, but roofed shed, where, by frequent sprinklings with water, the powder was reduced to a paste which was turned over and watered over and over again, for about nine weeks, because this odoriferous, heated, fermenting mass gave off steam, necessitating great care and even skill to the end that fermentation was neither so slow that a bulky product resulted, nor so rapid that it impaired dyeing properties. When fermentation had finally subsided, and the stiff paste sufficiently cooled, it was packed in casks ready for market. It was anciently estimated that nine parts, by weight, of woad leaves yielded one part, by weight, of finished pigment. As the woad plant was abundant and hardy, this was considered fair return, and in average years, profit, derived from it, was often greater than the value of the land on which it was grown. Therefore, many Europeans engaged in woad cultivation. However, the introduction of indigo from India, in the 16th century and of synthetic dyes in the latter part of the 19th century, reduced the use of woad to the vanishing point.

Although no valid trace of the woad plant has been found in Neolithic times, yet we have record that it was anciently grown in China, as the source of blue dye, and undoubtedly, it was also cultivated in Egypt, for some early botanical writers believe that the "robe of the ephod, all of blue" worn by the High Priest at Jerusalem, may have been dyed with woad, with which both the Egyptians and the Hebrews were familiar.

When, almost two thousand years ago, Julius Caesar's Roman army, after crossing the English Channel, invaded and occupied England for a few centuries, he found that the aboriginal inhabitants, an ancient Celtic race that had invaded the country after the Stone Age, and who were called "Picts," or painted people, had long punctured their skin with flint tools, and into the abrasion had rubbed anil of the woad plant, and thus formed various designs. In fact, it is suggested by Dr. Guest, in his "Origines Celticae," that the word "Britain," the oldest national name for England, is the Latinized form of "Brythen" a Celtic word, meaning "painted men."

In "Commentarii De Bello Gallico" (Book V) Caesar says: "All Britons stain themselves with woad (vitrum) which grows wild and produces a blue color which gives them a terrible appearance in battle." This statement is confirmed by Promonius Mela (Book II) who reports: "They dye their bodies with woad (vitrum) whether for ornament or any other reason is not known." About 70 A.D., Pliny the Elder, in his "Naturalis Historia" (Vol. XXII) undoubtedly refers to woad when he says: "There is a plant like plantain, called in Gaul glastum, with which the wives and daughters of Britain smear their bodies in certain ceremonies, and go naked, being of the color of Ethiopians." The word "glastum" used by Pliny, was derived from the Celtic word "glas" which meant blue. About 3 A.D., Ovid writes in his "Amorum" (Vol. II) of Britons as "Virides Brittanos," alluding to the dye with which they painted their bodies. Ovid also tells how ancient Teutons blackened their gray hair with woad, and may also have daubed their skin, as most primitive races were accustomed to painting their bodies, and, especially on the Continent, woad was available for that purpose. Lastly, Herodian (Book III) refers to ancient Britons as being ignorant of the use of clothes "but they mark their bodies with various figures of all kinds of animals, which is the reason they wear no clothes, for fear of hiding those figures." By contrast, the English poet Garth, referring, in 1715, to the extreme styles in dress of his day writes:

> "When dress was monstrous and fig leaves the mode
> And quality put on no paint but woad."

In the various early references to woad, as a pigment, Caesar describes it as blue, Pliny reports it as black, whereas Ovid refers to it as green. Therefore, many writers of botanical history may be excused if they express doubt as to whether or not the plant known to the ancient Celts and Britons, as the source of blue dye, was the plant we have so long known as woad. But modern scientific writers know that it was woad, for from the early Stone Age, it was known and used, in Britain and on the Continent, by the Celts, who had spread westward from the native habitat of the woad plant.

Soon after the Romans had been driven from Britain by the Picts, the Saxon tribes made the first of a series of invasions, and when they became settled, found that their consumption of woad, which they used to dye homespun cloth, was so extensive that it was necessary to bring additional shipments from their former home, across the North Sea. It was the Saxons who gave us the name "woad" and the earliest known use of it is found in an old English "Glossary of Names of Common Objects" which was prepared about 1000 A.D. Over a century elapsed before a form of this word appeared, but surviving records of that period are very few in number.

A British document, dated 1243 A.D., mentions duties payable on woad and another document, dated fifteen years later, is an agreement made between the citizens of Norwich, England, and woad merchants of Amiens, France, for even at that early period there appears to have been some sort of a woad "syndicate" in Europe. This was utilized, at a much later date, to combat the introduction of indigo from India, which, incidentally, was far superior to woad as a blue dye. In the light of modern business practise, this association, or syndicate, can be said to have been successful, for indigo is barely mentioned in English and French records until 1581 A.D., or nearly one hundred years after the all-water route to India had been discovered and was being served by British ships. In fact, the "syndicate" appears to have had sufficient influence, in some countries, to obtain laws prohibiting the use of woad's competitor, indigo, and for nearly three hundred years, the latter dye was forced to overcome tremendous opposition, before it finally replaced woad.

Writing in the 14th century, Geoffrey Chaucer used the Saxon name "wod" which apparently was the common form of his era, and is believed, by some etymologists, to have been derived from Wotan, the Teutonic god of war.

Prior to the much delayed introduction of indigo from India, woad was the only blue dye used in Western Europe and, at that time, woad was used more than all other dyes combined. It was the earliest plant to be cultivated for its pigment content. Throughout many centuries, woad was an important dye and especially in France and Saxony, the woad industry flourished and was so profitable that, as early as 1339 A.D., woad dyers were organized as a distinctive trade, "woad men" being limited to certain families who even had a traditional chant of their own. These men developed a dyeing technique which was passed from father to son, and it was early discovered that richness of the dyed shade achieved depended on the quality and quantity of woad and also on the frequency with which the cloth to be dyed was immersed in the dye bath. A fresh bath would give a deep black shade to almost every fiber fabric, and, as the solution grew weaker, it was, in turn, used to obtain blue, and finally, green. What, during the Middle

Ages, was known as Saxon green, was cloth dyed blue with woad and afterwards yellow with weld. If madder powder was added to the solution, at its weakest point, a deep purple shade was developed on the fabric, and this led to combinations with other natural dyestuffs to produce what, for that period, was a wide range of color, thus conferring upon woad the name of "Universal Dye of the Middle Ages." However, because of its complexity of use, woad, although providing fashionable colors, was never a cheap dye.

In medieval times, the principal woad growing districts of Europe were Saxony and Thuringia and, because the wealth of these States was largely based on the culture and sale of woad, this plant was called "the country's gold mine." Although it was true that, with more or less success, woad was cultivated in many other European countries, yet these two German States were about the only ones having sufficient quantity, over local needs, to export to Flanders and other important cloth-weaving centers of their times. During the 14th and 15th centuries, a considerable quantity of Saxon woad appears to have been sent to France and England and, as late as the 17th century, the latter country was an important importer, although, by that time, England was not only growing woad, but, in an effort to increase the scope of Indian export of raw products, was also importing indigo from India, much to the disgust of the local woad industry. The English government finally reversed itself on indigo protection, placing every imaginable obstacle in its way, even going so far as to pass laws forbidding the use of indigo because it "ruined the texture of cloth." France made many attempts to cultivate woad, and thus escaped the necessity of importing it. For many decades, French dyers were forbidden to use indigo, although all large cloth dyers realized that indigo possessed over ten times as much dye vitality as did woad.

During the Middle Ages, when woad was the chief blue dyestuff of Western Europe, supervision of the culture of dye plants was very strict. In certain countries, peasants were permitted to grow only a certain amount of woad and no dyer was permitted to buy up the "precious product" ahead of his colleagues. Peasants brought their woad crop to market where, prior to sale, it was subjected to test and often was marketed, by law, only to citizens of the same town.

During the 16th century, the woad plant, by that time widely cultivated in England, was subject to some peculiar laws enacted either by royal decree or Act of Parliament. At the behest of Edward VI, Parliament, in 1550 A.D., passed an Act which ordered "not any person shall dye any wool to be converted into cloth unless the same wool be perfectly woaded." It is recorded that Queen Elizabeth so disliked the odor which came from the fermentation of woad that she ordered that, during her "progresses" through the country, she "might not be driven out of the towns by the 'oade' infecting the air too near them." The sowing of woad plants was forbidden within five miles of any of the Queen's residences "whereby any offense may grow, from the noysome savour of the same."

After nearly twelve hundred years of supremacy as a blue dye in Western Europe, indigo finally replaced woad, but its reign was brief, for about two hundred years later, indigo was replaced by man-made synthetic dyes, which have left but little field for their historic predecessors.

SAFFRON

Saffron, the principal yellow dyestuff of the Greeks and Romans, was obtained from the pistils of the *Crocus sativus*, order of iridaceae, a plant which flowers in September or October and is quite distinct from the *Crocus vernis*, or common spring variety. The name of this plant is derived from the Arabic word za faran meaning yellow. Every place it was anciently introduced, it found a number of important uses, especially as a spice and as an ingredient in many complicated medical prescriptions, in which fields it was even more popular than when used as a dye. It found its way into Roman baths to perfume the air, and a contemporaneous record states that "the streets of Rome were strewn with saffron whenever the Emperor returned to the city with his army." In the early days of Greece, it was, at one time, the official color; Grecian women were particularly fond of saffron yellow for their clothes. The Egyptians were well acquainted with saffron, and Homer frequently refers to it in his writings.

This versatile plant was one of the chief trade commodities of the Middle Ages, for medieval cooks and physicians, as well as dyers, used the fine powder of the dried crocus. Monks discovered that by its use, in connection with an iron mordant, illuminated manuscript could, quite inexpensively, be made to resemble gold. For a short period, the city of Florence used the saffron blossom on its corporate coat of arms and, at the town of Semifonte, which was destroyed by Florentines in 1202 A.D., it was commonplace to raise badly needed funds by pledging a few pounds of saffron bulbs, rather than jewels or other personal property.

Saffron was long cultivated in Persia and, from there, found its way into China because of nomad Mongols. In Asia Minor, where it not only grew wild, but was crudely cultivated in districts which also were carpet-weaving centers, it was used to produce yellow shades on woven carpets, but as more than four thousand dried stigmata were required to produce a single ounce of dye-stuff, the use of saffron was abandoned as soon as other vegetable yellow dyes appeared. There is evidence that saffron was cultivated in Spain quite early, during the Arab occupation, but its introduction into Northern Europe is credited to returning Crusaders.

The city of Basle, Switzerland, also adopted the saffron blossom as part of its coat of arms, for this plant long provided wealth for the Swiss community. There was also a standing order from local merchants at the custom house of Basle, and saffron was eagerly purchased whenever a consignment arrived from Italy. The "Saffron War" of 1374 A.D. dramatically focused attention on the importance of this shrub, for as an incident of that occasion, a consignment of eight hundred pounds of saffron was seized as booty. This, and similar events, as well as the uncertainty and expense of medieval transportation, induced the cultivation of the saffron plant at Basle, especially as the widely traveled merchants of that city felt that they were familiar with its culture. Thus, by 1420 A.D., and for ten years thereafter, the cultivation and sale of saffron had become a profitable venture, and the City Council proclaimed "a course of events has arisen which should be a profitable venture for the merchants of Basle, namely the fact that many people, noble and common, have now begun to grow saffron which appears to be coming on well." In addition, the bailiffs of the city were ordered to "keep watch on our fields." The success of this saffron venture at Basle tempted other cities to similar efforts, for "they envied Basle merchants-growers, the City Council, and even the dean and chapter of the Basle cathedral who received tithes of every fruit borne by the earth." Many offers to sell saffron corms were declined, and strict local laws were enacted to prevent their export. However, this highly promising industry was of short duration, for scarcely ten years after its start, a succession of poor crops ended the sale of local saffron, but nevertheless, Basle remained an important center of saffron activities for Flanders, England and Germany.

A romantic story about saffron is told by Richard Hakluyt, the English geographer, in his "Voyages and Discoveries of the English Nation," written in 1660 A.D., when he states that saffron was first brought to England from Tripoli, during the reign of Edward III, by a pilgrim who "proposing to do good for his country, hid two saffron corms in his staff, made hollow for that purpose, all this at some danger to his safety, for had he been apprehended, he had died of the fact." This story may have been true, although Hakluyt was capable of romancing, for saffron, not native to England, is reputed to have been cultivated there as early as the 14th century, and was in production by farmers, who, until about 1768 A.D., were called crokers.

SAFFLOWER

The safflower plant is an isolated botanical species of the thistle, *Carthamus tinctorius*, an annual plant cultivated in Spain, Egypt and the Levant and belonging to the compositae. It provided a basic yellow dyestuff, similar to true saffron and was often used as a substitute for saffron, although these two plants are in no way related to each other. The safflower plant grows from two to three feet high. Its dye content consists of a water-soluble yellow and a water-insoluble red component, the two affording an orange hue. The dye is centered in the floret heads which are about one inch in diameter, and consist of many small flowers, growing at the end of numerous branches. When in full bloom, these florets are carefully picked by hand, and at once thoroughly dried in the sun, to obtain an orange colored fibrous mass, resembling saffron, or they are first kneaded in water in order to remove the fugitive yellow coloring matter, in favor of a reddish-yellow tint, and then pressed into cakes preliminary to packing in two hundred pound bales for market. Safflower owes its dyeing value to a small content of insoluble red coloring matter, which fixes the soluble yellow content to at least 30%. The versatility of this vegetable dye, especially when used with various acids and alkalis, is such that it will transmit yellow, orange, red, and shades of pink, but at best, it is a very weak dyeing agent, requiring from four ounces to one pound of extract per pound of fabric to be dyed, to obtain shades of color ranging from pink to crimson.

It is believed that the safflower plant originated in Southern Asia and, from almost prehistoric days, it has been cultivated in China, India, Persia and Egypt where its earliest use was for food and medicine. There are numerous references showing that, for many centuries, the Chinese used safflower to obtain rose, scarlet, purple and violet shades on silk and we are told that Egyptians used it to secure a brilliant scarlet on linen. From Egypt, the culture of safflower was gradually extended to the Mediterranean area long before the Roman era, but it seems to have been well known to ancient Greeks who used its seeds for a purgative and, for some reason or other, adopted safflower as their ancient official color. In medieval times, the safflower plant was cultivated in Italy, France, and Spain, and soon after the discovery of America, the Spanish Government directed that the safflower plant should be cultivated in Mexico and that portion of Spanish South America, which is now known as Venezuela and Colombia.

The cloth merchants of Northern Europe probably obtained safflower for dyeing purposes, either from Florence or Genoa, where, at the end of the 12th century, dyeing art had attained world-wide importance, and an English record of 1583 A.D. states: "Yee shall take one pound of saufleure and let it soke halfe a day."

The rapid disappearance of safflower, as a dye, was partly due to the fact that its several color shades, although bright and beautiful, were also very transitory. Moreover, the process of growing, gathering, and airing the dye-containing florets and of extracting the dye was not only time-consuming but also expensive. Today, safflower is rarely used in Europe, but large quantities are still cultivated for local use in India.

WELD

The weld or woald plant, *Reseda luteola*, is an annual growth, related to the mignonette family, and grows wild in waste places, in many parts of the world. It puts out long, narrow leaves of a lively green hue, from the midst of which arises a stem about as thick as a pipe stem. The whole plant, except the roots, serves for dyeing purposes. It was anciently known as "dyers' broom," "dyers' rocket," and "dyers' weed," remaining supreme for yellow pigments during many centuries. It was replaced only when old fustic was brought to Europe from the New World, in the 16th century.

The weld plant attains a height of about three feet and is pale brown in color. After cutting off at the root, followed by careful air-drying, it was sold in sheaves, like straw, but though bulky in appearance, yet each plant contained but a small quantity of yellow coloring matter, this being scattered throughout the entire plant, with the exception of the root. The seeds and upper portion of the plant are the richest in pigment.

Weld is of greater antiquity than any other yellow dye source, and was highly prized by the ancient Romans who restricted yellow to bridal garments. However, they used weld not only for that purpose, but also for the garments of the six Vestal Virgins who had consecrated their lives to the service of Vesta, in keeping alive the hearth fire which was never permitted to become extinguished. At the time of Julius Caesar, it was used by the Gauls and other nations north and west of the Alps and, in medieval times, was widely cultivated in France, Italy and Germany.

Brazilwood

Various leguminous trees, including the lima, sapan, and peachwood, are classed as brazilwood, or so-called soluble redwoods, to distinguish them from the barwood family of trees which only yield coloring matter dissolving with difficulty and only in boiling water. True brazilwood is known botanically as *Caesalpinia echinata*, and about nine varieties of this tree, have, at various times, and in various countries of the world, been used for dyestuff. These medium-size redwood trees were cut into small, regular sections, after which the units were rasped to coarse powder, moistened with water, and fermented for five or six weeks, in order to increase the coloring properties of the wood. Used with an aluminum mordant, this soluble redwood gave all kinds of fabrics, save silk, a bright red shade, and if bichromate of potash was used as a mordant, an agreeable purple-red could be obtained.

Prior to the 15th century, Europe obtained its brazilwood from India, Sumatra and Ceylon, by way of Venice, but with the discovery of an all-water route to India, brazilwood was brought direct to Lisbon in Portuguese ships. Although an oriental product, yet the word "brazil" is of Arabic origin, having its root in the word braza, meaning fiery red, which root also gave the word brazier, the dyers' furnace of red hot embers.

Brazilwood is frequently mentioned in medieval records, one of the earliest being a tax list at Ferrara, Italy in 1193 A.D., another tax list at Modena, Italy in 1306 A.D., and still another at Barcelona, Spain, then under Moorish rule, in 1280 A.D. After the year 1400 A.D., brazilwood is frequently mentioned in European records and, coincidental with the discovery of America, brazilwood became a common but valuable export for European traders, and even today is supplied almost exclusively by Brazil, with some less important shipments originating in Jamaica and Central America.

It is interesting to realize that the increased use of brazilwood in Europe and the contact with its two great sources of supply was the direct result of the discovery, by Vasco da Gama, in 1497, of an all-water route to India and the Far East, which made possible the delivery of Asiatic products direct to the port of Lisbon, without the expensive re-handling at the end of caravan and short-water hauls. Three years after da Gama had opened the Cape of Good Hope route, a Portuguese expedition, bound for India, lost its course and finally landed on the northeast coast of South America. Observing the large number of brazilwood trees, which were well known to them because of trade with India, they named the country "Terra de Brazil" and one of the earliest maps of the world, in the form of a globe, made about 1510 A.D., first used that name, but applied it to the entire northern portion of South America.

In one of his quaintly worded descriptions, Geoffrey Chaucer, the "Father of English Poetry" refers to brazilwood undoubtedly as a coloring matter, and his reference was to the same red dyewood so well known in Venice and Genoa during the Middle Ages. For many years in France, it was illegal to use brazilwood on soft fabrics, as it was thought that this dyestuff hardened the cloth.

Although still used, to a limited extent, both in wool dyeing and in calico printing, brazilwood has lost its ancient importance as a red dye, but it is occasionally used in conjunction with garacine, a logwood subsidiary, when a chocolate tint is desired.

Logwood

Logwood, also known as Campeachy wood because it was discovered by the Spaniards on the shores of the Bay of Campeachy, in Mexico, is, perhaps, the most important of all the older vegetable dyestuffs, and is one of the few that is still used on a large scale. It is peculiarly adapted for dyeing purple on wool, blue and black on cotton and wool, and black and violet on silk and is important today as a black cotton dye.

Logwood is derived from a fairly large tree which grows in tropical and sub-tropical America, known botanically as *Haematoxylon campechianum*, or the "blood-redwood from Campeachy," a tree distinguished by its peculiarly ribbed appearance. It has a thin, smooth bark of brilliant gray, sometimes yellowish, and seeds that have the taste of cloves and which the English call allspice. This tree has been scientifically cultivated in Jamaica since 1715. The wood is classed as a redwood, but the color-lake, as finally developed, is either black or blue, depending upon intensity. As it exists naturally in the wood fiber, the coloring matter is probably in the form of a glucoside, and freshly cut wood is colorless until exposed to oxidation by air, when the outside becomes a dark reddish-brown, whereas the inside becomes a pale yellow or orange. Logwood comes to the market in large blocks, each weighing up to 400 pounds. These are reduced to chips or to a paste, in which form it is aged by thoroughly wetting the mass and then heaping it in piles from four to six feet high. Because this process usually takes place indoors, fermentation is accelerated, and the logwood chips or paste requires constant attention, for, if fermentation proceeds too far, much of the coloring matter will be destroyed.

Logwood was introduced in Europe by the Spaniards, soon after the discovery of America, and was used in Spain early in the 16th century when it greatly expanded the art of dyeing and placed Spanish cloth merchants in favorable European competition. Logwood, however, does not appear to have been used as a dyestuff in England until the reign of Queen Elizabeth, and even then, its use was of short duration, for logwood met with stubborn opposition from the older school of dyers, who immediately joined in an effort to have it banned. In 1580, an Act of Parliament forbade it to be used for dyeing, and large quantities thereof were burned. Although persons, violating this law, were liable to imprisonment or the pillory, yet, until the reign of James I, many dyers, unable to meet Continental competition in any other way, apparently used logwood under other names. Not until one hundred years later, was the real value of logwood appreciated, and the harsh law repealed.

Barwood

Barwood is a hard resinous wood which grows in the equatorial regions of Western Africa and provides a color of deeper blue than do its important relatives, camwood and sander. Wood from all three trees is first ground and then boiled, because the color cannot be extracted if soaked as chips or, as was ancient practise, enclosed in bags suspended in the boiling water.

Camwood

This tree, which grows in Western Asia, is similar to barwood in its chemical structure, but is more abundant in coloring matter, producing a more intense red than does barwood. The coloring matter of camwood gives a harsh and disagreeable feel to cloth, because of its resinous qualities.

THE FUSTICS

Young, or Zante fustic, often called Venetian sumac, comes from the stem and larger branches of the smoke tree, *Rhus cotinus*, a shrub-size member of the cashew family, mentioned by Pliny in his "Natural History." It is a native of Asia and Southern Europe. The word fustic comes from the Arabic word "fustug" meaning small tree. Young fustic is also obtained from *Morus tinctoria* or Cuba wood, native to the West Indies, Central and South America. The wood of both varieties is hard, compact, and of reddish-orange color, thus providing the yellow tinge, peculiar to medieval scarlet. It was shipped to cloth dyers in the form of small-size logs or sticks, which, after rasping or grinding, were used for making dye, either in the form of chips or as a liquid. With selected mordants, young fustic imparted colors to cotton and wool fabrics, varying from bright yellow, orange and yellow to brown or dark olive, but it was also used with logwood, to obtain a dead-black shade. As a dye, young fustic is of much greater antiquity than the now much more important old fustic, and incidentally, the use of the words "young" and "old" is misleading, since there is no botanical connection between them. Young fustic has almost completely disappeared as a source of dye, for it has little permanency and, in later years, was seldom used alone.

Old fustic, the "bois jaune" of French dye-masters, even today an important natural dyestuff, is the golden-yellow wood of a large tree of the mulberry family, botanically known as *Chlorophora tinctoria*, which grows wild in the West Indies and tropical America, and was first brought to Europe by the Spaniards, about 1510 A.D. The best old fustic is grown in Cuba, Jamaica and Brazil, where trees often reach a height of more than sixty feet, and, when felled for dye-making purposes, is exported in the form of large blocks, which are brown outside, brownish-yellow inside, and free from bark. The dyes, obtained from this tree, were principally used for woolen fabrics, to which it imparted shades of yellow, varying from old gold to lemon yellow. Today, old fustic is used as an extract, in combination with logwood, for dyeing wool and cotton various shades of brown or olive.

Orseille

Scientists assume that about two million years ago, some fungi accidentally became intimately associated with algae of the blue or blue-green variety, and, as this association evidently appealed to nature as being beneficial to both of these organisms, it was continued. This is an explanation of the origin of all lichens, of which orseille, also known as archil and many other names, is a prominent member. The orseille growth, which Linnaeus named "Lichen Roccella tinctoria" which he knew as growing on the rocks of many Mediterranean islands, and particularly the Canary Islands, grows upright, partly with single, partly with double stems, each of which is about two inches high, and when the plant is mature these stems are crowned with formations either button-like round in shape, or flat, varying in gray color from light to dark. It is from these heads that a dark red paste is made.

In the remote past, peasant dyers of the Near Fast and Mediterranean area used orseille for simple coloring purposes, when a purple shade was desired. Both Theophrastus and Dioscorides describe how this lichen grew on the rocks of various Mediterranean islands, notably Crete and Candia, both adding the statement that it was used for dyeing wool because "when fresh, its color was so beautiful that it even excelled the ancient purple of Tyre." Pliny records that dyers of his day also "gave ground" (i.e., preliminary cloth dye bath) to those more costly cloths which were afterwards to be dyed with Tyrian purple. Although orseille as a dyeing agent had been used for centuries, in the Near East, and was not unknown to the ancient Greeks, yet many European dyers of the Middle Ages thought that it was discovered about 1300 A.D. by a Florentine dye trader named Federigo. He not only carried on an extensive dye business in saffron, indigo, and kermes with the Near East, but had widely traveled in the Levant and, unquestionably, was the first European to proclaim the merits of orseille as a purple dye. Federigo promptly perpetuated that event by adopting the patronym of Oricelli, which, almost at once, became an influential Florentine family, many of its members having been distinguished scholars and statesmen.

Orseille had its first European use at Florence and, for many years, dyers of that city, and also those of its rivals, Genoa and Venice, imported orseille from the Near East solely for their own use. Because of the good reputation of

Italian orseille-dyed cloth, they gradually built up a foreign trade in orseille dye paste and ultimately supplied it to English, Flemish and German cloth merchants. Apparently, for many years, few French dyers used orseille, for it is not mentioned in any of the older French dyeing manuals. This trade in orseille dye paste was held for nearly four hundred years, or until 1703 A.D. when the orseille lichen was discovered on the comparatively near-by Canary Islands. So avid were English and Dutch cloth dyers to obtain orseille from this new source that an English traveler, writing in 1704 A.D., expressed surprise that Europeans, "immediately upon their arrival on these (Canary) islands, sought after this lichen with as much energy and skill, as they did after gold in Spanish America, though they were not so well acquainted with the former as with the latter before the discovery of these new islands." Unlike the Florentine cloth dyers, their English rivals would not purchase raw orseille, but preferred it in paste form as previously supplied from Florence. As the preparation of this paste was a closely guarded secret, the discovery of orseille on the Canary Islands, although welcome as an independent source of supply, yet offered but little advantage until, fortunately, the paste formula was divulged to them by an Italian cloth dyer of London, who, in former years, had carried on his trade at Florence.

Orseille is described by George Eliot in "Romola" as "a little lichen which grows on a rock, and, having drunk a great deal of light into its little stems and button-heads, will give it out again as a reddish-purple dye, very grateful to the eye."

Orseille, or archil, was the most important lichen dye derivative of ancient and medieval cloth dyers, but other lichens were sometimes also used, the most prominent of these being lacmus which was known to have been brought from Norway, as early as 1316 A.D., for use in England and Flanders. Later in the Middle Ages coloring matter was also obtained from lichens *Roccella fuciformis*, and *Variolaria orcina* found both in Norway and on Mediterranean islands. Dye lichens have also been found on the coasts of Lower California, where attempts have been made to develop an American orseille industry.

CUDBEAR

This fine powder of lilac color is derived from lichens, especially *Lecanora tartarea.*

ANNATTO

Annatto was obtained from the pulpy portion of the seeds grown by the plant *Bixa orellana* and was originally imported into the Near East and Europe from India, where, although annatto has always been extremely fugitive to light, it had been used, for centuries, as an orange-red dye. After seeds and pulp had been removed from mature fruit, the residue was macerated in water, the product strained, and the coloring matter, which had fallen to the bottom, collected, dried in the sun, and formed into cakes. Some years after the discovery of America, the annatto plant was found in Central America and Brazil, but by that time, its use as a dye had been largely discontinued.

Turmeric

Turmeric, which derives its name from the Latin words "terra merita," is botanically known as *Curcuma longa*, a native of India and China. The yellow dye was obtained from its roots and when these were of good quality, were hard, having a dull, waxy, resinous fracture. Turmeric roots, when ground, produced a bright yellow powder that was anciently used for dyeing colors, varying from light yellow to orange. Turmeric was very popular, for many centuries, not only in Asia, but also in Greece and Rome, where it was used for dyeing silk and wool. Today, it is found only in India and Asia Minor where its use is largely restricted to carpet dyeing.

Cutch

Cutch obtained its name from a native State of India where it has been used as a brown dye for over two thousand years, because it was exceedingly fast to light and acid. It is an extract from the wood *Acacia catechu* and the Mimosa trees. The best cutch appears on the market in the form of square blocks weighing from 28 to 56 pounds each; the second quality cutch is in tablets weighing from one to two pounds each, whereas the third, or cheapest quality appears as large cubes.

Cutch wood is gathered when the acacia trees are still green and quite full of sap, and have reached a diameter of about one foot. The bark is first stripped off, the timber chopped into sections, covered with water, and boiled in a large cauldron. When the extract, which solidifies as it cools, becomes thick, it is molded into forms and when so prepared, is a dark brown, almost black substance.

GAMBIR

Gambir, or Gambler which is chemically similar to cutch, is an Asiatic plant whose leaves and young twigs were boiled to secure a pale-brown or yellow dyestuff which was locally used.

QUERCITRON

Quercitron, or Quercetin, is a crystalline yellow substance, obtained mainly from the inner bark of *Quercus nigra*, or black oak, and sometimes from the inner bark of *Quercus citrina*, erroneously called yellow, or lemon-colored oak, both of which yield a brownish-yellow color, after the outer bark has been removed. The inner bark is reduced to powder by grinding and provides as much coloring matter as ten parts of weld or four parts of old fustic. Quercitron occurs in small quantities in both apple tree bark and horse-chestnut leaves.

The Animal Dyes

Tyrian Purple

Tyrian purple was the most highly prized and expensive dyestuff of ancient times and thus, it is fitting that its discovery should have been the theme of a fascinating legend which was long perpetuated on coins of Tyre. It is related that a sheep dog of Greece's greatest hero, Hercules, when attempting to bite into a shellfish, stained his jaws bright red, and noting this, the dog's master at once realized the significance of what had occurred, and ordered a gown to be dyed with the newly found color. Historians and archeologists agree that Tyrian purple dye, for home consumption only, was first made in Crete, as early as 1600 B.C., but its distribution, by seafaring Phoenicians, in the dawn of Mediterranean history, brought prosperity to Tyre as early as 1500 B.C. and continued to do so, in more or less volume, until the final conquest of Tyre by the Arabs in 638 A.D., having survived the political decline of Phoenicia, without itself being wholly destroyed. At the peak of their operations in Tyrian purple, the Phoenicians maintained trading centers and dye factories all around the Mediterranean coasts and even on the west coast of Africa, making and bartering purple dye at ancient trading posts which are, today, familiar to us as Cadiz, Carthage, Tarentum, Palermo and Marseilles, with the far-western Atlantic post in present day Morocco.

This brilliant and non-fading dyestuff is the only dye described, at great length, by Pliny and other ancient writers, who relate that the Phoenicians ceaselessly searched for pupura shellfish on every coast accessible to their small craft and, when found in quantities, as at Tarentum, promptly established trading stations and subsidiary factories which they absolutely controlled, although in the later days of the Greeks and Romans, purple dyes were pretty generally, independently made and sold throughout Western Asia and the Mediterranean area.

Pliny states that the method of preparing the dye was very complicated, varying with each of the four or five different shellfish harvested for that purpose. In the case of the larger mollusk, *Murex brandaris*, or "Turk's blood," found in ancient times on the Italian coast, near Tarentum, but a single drop of glandular mucous was extracted from a gland, adjacent to the respiratory cavity of each mollusk. This fluid first appears white, but, upon exposure, becomes yellow-green, finally reaching the last and permanent natural color of either violet or reddish-purple.

The smaller shellfish. *Murex trunculus*, found along the Phoenician coast, near Tyre and Sidon, were crushed, and the crushed mass was salted for three days, following which, the whole batch was boiled for about ten days, or until the dye showed the anticipated scarlet hue. Pliny said that the best season for catching these shellfish was either autumn or winter, as during summer breeding, they remained in deep water.

It is necessary to bear in mind that only for the past two or three hundred years has the word "purple" meant a definite shade, with but one or two variations. In ancient days, Tyrian purple included many shades such as deep red, blue, violet, black and even green tints in the otherwise purple basic shade. The possibility of arriving at different tones, resulting from chemical changes, when the murex juice was exposed to light, made it necessary to interrupt dye processing at various points, and even to dilute the basic dye with water, honey, or orseille and thus obtain blue, red, or green, each in a variety of shades, of which the two most prized colors were purple and red. The earlier references to purple undoubtedly refer to a very costly bright-red purple which was fixed to retain the color of coagulated blood, but which, when viewed directly, appeared to be black, and a lustrous red, when viewed obliquely. One shade of purple, frequently noted by both Pliny and Josephus, "resembled the color of the sea, the air, and a clear sky" and was therefore called blue, a color long considered a variant of purple and much used by the Hebrews in various tabernacle decorations. In fact, blue and purple, in Biblical literature, are interchangeable terms, and were so closely associated with conceptions of sovereignty that, when a monarch bestowed a purple robe, as did the King of Persia upon Mordecai and the King of Babylon upon Daniel, the act almost automatically conferred a position of power.

Purple was not the oldest color known to man, but, from the day Tyrian purple was first made in those ancient factories at Tyre, Tarentum and Palermo, it was very rare and costly because it was necessary to capture and use so many thousands of shellfish for so little dyestuff that it had always been a color of distinction, long restricted to regal and ecclesiastical uses. The Kings of Media, the royal houses of Persia, Babylon and Syria, all wore purple, thus creating a large demand for this Phoenician dye. Purple and white was the royal robe in which at Susa, his capitol, Darius of Persia advanced to meet Alexander of Macedon, and the absence of purple in the attire of his youthful Greek conqueror quite amazed the defeated Oriental ruler. Thereafter, purple was not only the color of robes of Greek generals and of those "attributed to the gods," but the Greeks having "humanized" their gods, an idea that had not previously entered the mind of man, dressed them also in purple. It was said that when the Greek army captured Susa about 330 B.C., they found "purple robes to the value of 5000 talents"—about $7,000,000— "some of which had been stored for two hundred years" without losing any original lustre. Purple was the color of the sails on Cleopatra's barge, on which the faithless Marc Antony fled from Actium. Both Julius and Augustus Caesar decreed that in the Roman Empire, none but the Emperor and his household might wear purple, and this remained the law until Tiberius annulled it. In the Eastern Roman Empire, the heir to the throne at Byzantium bore the proud name "Porphyro-Genitus," born to the purple.

Violet, or Tarentum purple was the popular Roman shade, and Martial records that, about 50 A.D., if a purple robe could be purchased at all, it would cost $550. In the reign of Diocletian, a palace purchase order listed 328 grams of silk, dyed Tyrian purple, at a cost of $875. The early Christians abhorred purple as evidence of luxury, and as early as 200 A.D., St. Clement, Bishop of Alexandria, wrote "I am ashamed to see so much treasure expended to cover shame."

Kermes

If we omit lac dyes of India, the oldest of all insect dyestuffs was kermes, an oriental shield louse which lived on the leaves and stems of low, shrubby trees, having prickly leaves, principally the holm oak, *Quercus ilex*, and the shrub oak, *Quercus coccifera*. This insect has been variously called *Coccus arborum*, and *Coccus ilicis*, but, since antiquity, it has been known as *kermes*, an Armenian word, meaning "little worm." This word has been traced by etymologists from the Sanskrit *krimija*, the ancient Persian *krmi*, and the Arabic *qirmiz*, all of which mean "worm begotten." The insect, however, was "adventitious and not natural to the tree" and possibly not every oak or holly shrub was a kermes haven, but these insects were sufficiently numerous and were distributed over a large area, so that, for nearly 3000 years, they provided the scarlet dye widely used from ancient times until the Middle Ages. It was frequently mentioned by Hebrew and Arabian writers and it is interesting to note that, unlike their Greek, Roman and European successors, these ancient people of the Near East were aware of the animal origin of kermes, whereas Greeks, Romans, and Europeans, until the 17th century, believed kermes to be a vegetable growth, defining it as "the proper fruit of the (oak) tree" (acorn) and therefore classed it as coccus. In the first century A.D., Dioscorides said that kermes were gathered from *Coccus baphike*, a small, gnarled shrub on which the berries are clustered, and are picked from the bush and collected. Pausanias, a Greek geographer of the second century A.D., believed that the berry contained a live insect, while other writers thought it was the "scurf or scale of the tree" (*scabies fructicis*). But these ancient observers should not be too severely chided for misstatements, as even more modern writers compared kermes with "lentils and peas," and possibly, the first accurate European description of the insect appeared in 1551 A.D. One hundred more years were to pass before the insect origin of kermes was accepted as the result of researches by a Dutch chemist, about 1640 A.D., who discovered the similarity of kermes to cochineal.

Many dye chemists think that kermes was first discovered in Palestine by the Phoenicians, and Theophrastus, the Greek botanist, calls it phoinikos, which itself suggests Phoenicia. Aside from this possible link, there are still other grounds for ascribing the discovery of kermes-dyeing to the Phoenicians, whose experience in the use of murex (purple) dyes undoubtedly had developed technical ability in the field of dyestuffs. From several references in the Old Testament, the Phoenician origin of scarlet dyeing may be implied. In addition, there are no records of Phoenician use of vegetable dyestuffs, although they were a trading nation.

As this coccus is mentioned by Moses and other Hebrew writers, it is evident that kermes must have been known in Egypt and some of the remote countries of the East, at a very early period, for there is no record that kermes is or was indigenous to Egypt and it certainly was not the discovery of the small, wandering Hebrew tribes. Moses mentions kermes under the names "tola" or "tola shami," the prefix meaning "worm" and the suffix "bright (or double or deep) red dye." All ancient writers translate the Hebrew word tola to mean kermes, "a deep red bright dye." Also, when in the wilderness, Moses needed a scarlet dye to ornament the Tabernacle, he could have secured it only in Egypt, for Moses, at that time, had never been in Palestine. Tola was the ancient Phoenician name for scarlet, used by both the Hebrews and Assyrians, and evidently also by the Syrians, for it is used by the Syrian translator Isaiah for the first time in recorded history, when, it appears in the promise of Jehovah: "Though your sins be as scarlet (tola) they shall be as white as snow; though they be red as crimson (tola shami), they shall be as wool." There is also an interesting chemical conjecture involved in this early use of red dye on animal fiber. Could these ancient craftsmen have been aware that animal (either worm or shellfish) dyes are brightest and fastest on animal, rather than vegetable fibers, and was this statement so phrased as to suggest the potency of divine promise? During the long Babylonian captivity, the Hebrews dropped the word tola, and red kermes dyes were known by the Aramaic name of zehori.

Kermes was known in Homer's day, or during the 8th century, B.C., and trade in kermes between Greece and Sardis, capital of Lydia, was extensive.

According to such an experienced botanist as Dioscorides, "kermes were collected in Galatia, Armenia" and "Asia Minor," possibly, a careless recording by a hurried man, for the author of this statement first mentions two Asiatic districts, and then Asia, in general, as the most important districts where kermes was found. We are informed by Pliny the Elder that kermes were collected "from Asia and Africa," from Attica, Galatia, Cilicia and also from Lusitania and Sardinia, but those produced by the last mentioned places were of the least value.

Many modern authorities believe that the art of using kermes, to dye a beautiful red color, was discovered in the

Orient, at a very early date and that it was so improved and refined as to excel even the famed Tyrian purple. Thus, kermes played a large part in causing the proper purple not only to be abandoned, but also to become one of the "lost arts" until rediscovered about one hundred years ago. The costly red dyes, so highly praised by Hebrew writers, were, in the opinion of many, made from kermes. There is proof in the writing of "Vita Aurelian" by Vopiscus, where he informs his readers that about 275 A.D., "Hormisdas, King of Persia, sent to the Emperor Aurelian, besides other articles of great value, a red woolen cloak of native production which was of a much costlier and brighter purple-red than any that had ever been seen in the Roman Empire, and in comparison to which all the other purple cloths worn by the Emperor and the ladies of the Court appeared dull and faded." Textile chemists are of the opinion that this cloth, which was of a "beautiful purple-red color," was dyed with kermes, and not with the dye of the murex, but this fact was not likely to have occurred to Romans of that era, who were acquainted only with the latter purple dye, and who had less experience in peaceful arts and crafts than in war-like pursuits. Aurelian, and some of his immediate successors, ordered the most experienced Roman cloth dyers to search in India for this new and brilliant dyestuff, but these trade ambassadors returned without success, offering a vague report that the most admired Persian dye was produced from a local plant, which is presumed to be the madder, and which was already known at Rome. It is historically true that kermes was also known to Roman cloth dyers of that day, but their knowledge of the art of dyeing was so meagre, that they merely used kermes as a ground for the purple murex dyes and thus, it did not seem reasonable that any Roman dyer of intelligence would admit that India could produce a purple more beautiful than was their own murex purple. On the other hand, it is quite likely that these same Roman cloth dyers were not acquainted with the qualities of kermes which had long been known in both India and Persia.

From the tariff schedules of Diocletian, proclaimed about 300 A.D., we learn that scarlet wool of Nicae, in Bithynia, was the most valuable of dyed wools, excepting purple. Scarlet dyes were so highly prized by the Romans that they frequently formed part of the tribute exacted from a conquered nation, and thus, after their subjugation by Rome, the people of Spain were compelled to pay one-half of their tribute in kermes. This old Roman custom was followed during the 13th and 14th centuries, when many feudal landlords and monasteries, in Europe, accepted a portion of their rent and tithes, due to them from peasants, in the form of kermes, and dyestuffs were always acceptable at the monasteries because of their activities in weaving.

Kermes was the only dye of insect origin used by ancient and medieval dyers of Western Asia, and, although expensive, was less costly than purple, which was the color then preferred by cloth dyers, and their rich and noble clients, and which usually had a tinge of red, and not a few contemporaneous chroniclers record that kermes was used to "give it ground." As a matter of fact, purple was often mixed with kermes to such an extent that many Hebrew, Syrian, and other ancient writers speak of scarlet when purple was intended, and vice versa.

When, about the beginning of the Middle Ages, the principal trade center for Kermes and other oriental dyes was Venice, the nobles of that city always wore black, as the symbol of their republican status, yet Venetian cloth merchants supplied all Europe with scarlet clothing, for the best European dyeing was the product of the craftsmen of Venice. Kermes was exclusively used to give a brilliant red color in a various range of shades, and from these experiments in color are derived many of the words now in use to define the many shades of red. Kermes meant "little worm" and the Latin equivalent is *vermiculus*, from which root comes our current word vermilion. Pointing almost directly to this parent word, kermes, is our color-word carmine, whereas another color-word, crimson, passed through a preliminary Latin form of *carmesinus*. Our color-word scarlet, has an even more interesting origin, for English and American lexicographers have decided that although its Persian root is "sakirlat," which means red color, yet, it is derived from *carnis*, the Latin word for flesh and incorporated in Latin dyeing nomenclature as *scarlatum*, meaning flesh colored, and later, as *scarlatinus*, a pale red which was produced by mixing kermes with white, and finally *scarlato*, which defined the full, deep red of the kermes pigments. The name scarlet first occurs in Central Europe during the 11th century.

Dyestuffs, made from kermes, seem to have been known in Germany as early as the 12th century, for among the native products, sent by Henry the Lion, as presents to the Greek Emperor, we find mention of scarlato. Early English cloth dyers undoubtedly purchased kermes from either France or Spain, for this insect was indigenous to both countries. Arabian conquerors, who from antiquity had been familiar with kermes which had originated in Armenia, recognized the kermes insect in Spain and harvested it not only for local dye purposes, but also as an article of commerce. Because of this, some cloth merchants knew it as *alkerkes*, in the same way as they knew cotton as *alkotan*, although some historians assert that the Moors knew kermes as *vermes*. Here, it is worth noting that, as early as 1200 A.D., it was known to cloth dyers of both Asia and the Mediterranean area, that dyestuffs from animal sources, that is, insects and shellfish, gave more brilliant color with increased fastness to light and water, than did dyes obtained from vegetable sources. It was later to be known that when these animal dyestuffs were used on animal fibers, such as wool, silk, vicuna, and similar fabric sources, the highest dye value of that period could be obtained, and it was on wool and silk that kermes was used.

At a later time, when Roman dyers knew that the beautiful oriental kermes dye was not a true purple, it was no longer classed as such but was considered to be a new dye, and was called carmesinus, and cloth dyed with it was sold as scarlata. Vossius quotes several writers who use variations of this word, the oldest of whom were Caesarius, who lived about 1227 A.D., and Matthew of Paris, who, writing in 1245 A.D., referred to this word as being used in 1134 A.D. But even earlier than that, Emperor Henry III, about 1050 in conferring a dignity upon the Count of Cleves, did so upon condition that the latter deliver "tres pannos scarlatinos anglicanos" —three pieces of scarlet cloth made of English wool. This word is found in the writings of Petrus Mauritius, the navigator, whose name was given to the island of Mauritius, and who died in 1157 A.D. it also occurs about 1175 A.D. in anecdotes of Arnold, the first abbot of Lubeck.

With the decline of purple dyeing in the Mediterranean area, the importance of scarlet increased, especially so, as its brilliance was fully equal to purple. In 1467 A.D., about fourteen years after Constantinople fell, and the Turks closed the Mediterranean Sea to all but their own commerce, Pope Paul II ordered that "scarlet is the cardinals' color."

In the "Hierozoicon" of Samuel Bochart, the 17th century French historian, appear passages from manuscripts of Arabian authors which undoubtedly refer to kermes shipped to Venice in the form of round reddish-brown balls about the size of a pea. Each so-called pea had a tiny hole filed with dark, crumb-like particles, which, when pounded, yielded a red powder, soluble in alcohol and water. These grains were then shipped in tubs or crates, and owing to the fact that kermes was supposed to consist of grains, dyeing with kermes was known as "grain" or "ingrain" dyeing and this reference to "grain" is historically interesting. In Latin, the word "granum" means seed, and the small kermes particles were sometimes called by that name, because of their similarity to small seeds. Thus, grain indicates the red coloring matter in the dye of kermes and is a name of great antiquity. Ingrain, therefore, is literally an abbreviation for "dyed in grain," and Shakespeare so uses this word in "Twelfth Night" when one of the characters in this play exclaims "'Tis ingrain, Sir! 'twill endure wind and weather."

A very old Latin account records that there was a variety of kermes in Sicily which resembled a "small snail, and which was collected by women with their mouths." The comparison to small snails may not seem inconsistent, but the method employed for gathering hardly lends itself to credulity, for even far more ancient Latin writers had noted that "when chewed, these grains have a somewhat bitterish taste, and communicate to spittle a brownish-red color." Even at a more ancient time, the harvesting was done by the women of Asia, Greece, and other eastern Mediterranean countries who, as well as French women in more modern times, permitted their finger nails to reach a more than normal length in order to assist them in their work. Usually one woman could pick about two pounds of insects a day, the work being done by candle light, before daybreak, for at that time the dew had not yet evaporated and the thorny holly or oak leaves were still soft.

Another variety of kermes, not mentioned by the ancients, was found on the roots of some plants. Those insects, which entomologists call *Margarodes polonicus*, were collected for dye purposes in both Germany and Poland as early as the 12th century. Old ecclesiastical records relate that in the 13th century, several German monasteries caused their neighbors to collect them as a contribution and those who could not produce any were solicited to donate money. As it was the usual custom to begin this harvest with religious ceremonies at noon, on the feast day of St. John, it was sometimes called "St. John's blood," and some contemporaneous writers suggest that the clergy favored this name, as it suggested the religious use of the annual donation. As German monks and nuns of that era engaged in weaving, they could employ "St. John's blood" to a very good purpose. At one monastery, excess dyestuff "was sold by the Archbishop of Arles to the Jews of the diocese of St. Chamas." When the practise of paying tithes "in kind" was discontinued, all efforts to produce "St. John's blood" ceased. It is not recorded how long its use by religious houses continued, but we know that it lasted longer in Poland than in any other country, for from that country, as late as 1792, a considerable quantity was sent every year to Venice. This type of kermes did not find wide acceptance for several reasons: first, root kermes contained less coloring matter than did kermes imported from France and Spain; second, its collection was laborious as well as tedious, for the kermes "acorns" were not located until after each plant had been lifted out by its roots, the insects removed, and the roots cleaned; third, although the shrub was carefully replanted, yet digging up the root often killed the plant and thus required replacements, and fourth, after it ceased to be paid "in natura" to the monasteries, it was more expensive than kermes brought in from France and Spain.

COCHINEAL

Cochineal, *Coccus cacti*, is a natural, rich, crimson dyestuff, obtained from an insect of the same name which feeds on Cactus nopalea cochinellifera, and was used for the production of scarlet, crimson and other red tints, differing from kermes, *Coccus ilicis*, chiefly in shape. When introduced into Europe, it was largely instrumental in replacing age-old kermes for the dyeing of red tints on silk and wool fabrics. This popular dyestuff, one of the first to come from the new world, was introduced into Europe from its home in Mexico, or New Spain, as the country was then called, by Spanish invaders who reported that it had long been known in the new world as nochezli, and was for many centuries, prior to the invasion of 1518 A.D., not only in common use as a house paint and textile dye, but as in the case of kermes, in ancient Rome, it had been used by the Aztecs as a medium of tribute, which is a sure index of its value. The Spaniards related how from 1480 A.D. until the arrival of the Spanish invaders, Montezuma had exacted from the State of Huaxyacas (modern Oaxaca) a yearly tribute of twenty sacks of cochineal, for this State was the main source of cochineal supply, until under Spanish rule, Dominican Fathers appreciably widened the cultivation area.

There were two principal forms of cochineal, one being the so-called silver variety, which displayed a grayish-red color with the furrows of its body covered with a fine down or white bloom; the other, the so-called black variety, was dark reddish-brown, and destitute of any down. As cochineal insects require but three months to mature, a crop of about two hundred pounds per acre was usually collected during May, July, and October, but with the approach of the rainy season, branches of cactus plants, loaded with young insects, were cut off and carried indoors to prevent insects from being destroyed by stormy weather. When the harvest started, the insects were carefully, brushed from the cactus plants, upon which they feed, into bags, or small wooden bowls with sharp rims, squirrel tails being used for this purpose. Then, they were killed either by being packed in baskets for immersion in scalding water, or by heated ovens, or even by long exposure to the hot sun. This latter method, as with kermes, is considered as producing a superior quality of dye. Under either of these drying methods, the insect bursts open and turns rust-red, indicating that the process was completed. The dried insects had the shape of irregular, fluted or concave, current-like grains of a reddish-brown or violet-brown color. About 50,000 of these are required to weigh two pounds, and it has been estimated that, on the average, it requires 70,000 dried insects to produce one pound of cochineal dye.

The Spanish word for wood or shield louse is *cochinilla*, a diminutive of *cochina*, or small female hog. This term was applied to the insect because of some fancied resemblance in shape. When the Spaniards came to Mexico, in 1518 A.D., they promptly transferred this name to the dye-bearing insect, which had long been used by the Aztecs, because it resembled the wood louse of their native Spain, which, however, had not been used as a dye source. As the Spanish had long used kermes, they did not fail to detect the superiority of cochineal and to note that this insect was far more prolific. They reported these facts, whereupon, as early as 1523 A.D., the Spanish King, Charles V, directed Cortez to inform him at once "whether what has been reported is true that kermes were to be found in abundance in New Spain and, if so, could they be sent with advantage to Spain." This practical monarch added "should this information be true, pay attention to it, and cause as much as possible to be collected with diligence." So Cortez promptly initiated the Aztec system of tribute and very shortly afterwards, the export of cochineal began. Spain reserved a monopoly in its sale, and before long cochineal entirely displaced kermes, in Europe, which for nearly fifteen hundred years had been used as a red dye by European and Asiatic cloth merchants. Cochineal was not only superior to kermes as a dye, but it could be produced in abundance throughout the year, for it was cultivated on large Mexican plantations, and at a cost, which, if not as low as that asked for kermes, yet was moderate considering its quality. With the intention of preserving its profitable monopoly, the Spanish Government early forbade the export of live cochineal insects, and to this measure is attributed the fact that for many years, the origin of this important dye remained a mystery.

The earliest European mention of cochineal appears in the writings of the Spanish historian, Lopez de Gomara who, in 1525 A.D., described it as "an excresence of the nopal," which would make it an exudence of gum or sap, obtained either naturally or as the result of a bruise. Because of this early error, the true source of cochineal was not known, which greatly pleased Spanish authorities, who, from then on, closely guarded the truth. That cochineal soon became an important article of Spanish commerce is proven by the fact that Guicciardini, who died in 1540 A.D., several times mentioned that cochineal was "one of the articles obtained in Spain by cloth merchants of Antwerp" and, in 1556 A.D., the Spanish historian Oviedo mentions "an excellent dye which Mexicans prepared from the fruit of the cactus which we call cochinellifera, forming it into small cakes." However, when questioned about this by eager merchants, he acknowledged that he had "received no authentic account on this subject," but the cakes he referred to, undoubtedly were cochineal, for Spaniards, who had been in Mexico, had also reported them. Even as late as 1651 A.D., Hernandez, the Spanish author of "Study of the Natural History of New Spain," held the view that cochineal was a vegetable product, and it was not until fifteen years later that Plumier, a French naturalist, having discovered similar insects on the island of San Domingo, demonstrated that cochineal, like kermes, was of insect origin, and another naturalist advanced the theory that it might be a species of ladybird and even published a paper to support this view. In 1633 A.D., Johnson, an English writer, stated "upon this plant, in some parts of the West Indies, grow certain excrescences which, in course of time, turn into insects."

Cochineal appears to have been unknown as a scarlet dye in Italy, as late as 1548 A.D., and its general use in England was even later. About 1643 A.D., Kepler, a citizen of Flanders established the first English dye house for scarlet at Bowe, near London, and marketed his product as "Bowe-Dye." In 1667 A.D., King Charles I, with promise of large salary, invited another Flemish citizen, Brewer, to settle in England. To these craftsmen is credited the perfection of cochineal dyeing in that country.

Until about 1725 A.D., it was also commonly believed that cochineal was the seed of some American, tropical plant, although in 1672 A. D., an English botanist, Dr. Lister, had suggested that "it might be a sort of kermes." In 1703 A.D., two years after the ladybird theory had been published, a Dutch scientist, with the aid of a microscope, finally discovered some of the characteristics of cochineal and came pretty close to its true nature. It was this Dutch scientist who tested cochineal grains, known to be over a century old, finding their dyeing potency in no way diminished. It was Linnaeus, who, in the tenth edition of "Systema Naturae," gave cochineal its real place in the scientific order, for one of his pupils, who had visited Mexico, had sent him a cactus plant, complete with insects, the package arriving in 1756 A.D., at a time when Linnaeus was absent from home. His impractical gardener proceeded to clean the "vermin" from the plant, doing so well that but a solitary survivor remained for examination.

The Spaniards were content to center cochineal culture in Mexico for nearly four hundred years, but when rumbles of political unrest began to appear, in the latter part of the 18th century, a Spanish official at Vera Cruz sent live cochineal insects on cactus plants to Spain, where they were transplanted with only partial success, at Cadiz and Malaga. Similar efforts on Mediterranean Islands, and eventually in the Near East were more successful. Even earlier than this, some French dye chemists attempted to raise cochineal insects at Port au Prince, San Domingo, but this effort was unsuccessful as was also a later English attempt to propagate cochineal in India. Both the French and English attempts were directed to break the long-held Spanish monopoly in cochineal.

The decline of cochineal began in 1858 A.D., with the introduction of aniline red, but for the next two decades the new dye merely depressed cochineal prices. When azo dyes were introduced, about 1880 A.D., the use of cochineal for dyeing purposes practically ceased.

Lac

Lac dye is a resinous incrustation, produced by the insect *Coccus* (*Tachardia*) *lacca*, native of Indo-China, Siam and Southern India, and is closely allied to kermes of Asia, and cochineal of America. This insect, which usually appears in November, lives on the twigs and fleshy, young branches of the banyan and other Asiatic trees of the genus *ficus*, which grow in Bengal, Burma and Siam, but is particularly partial to the fig tree. The word lac is derived from the Sanskrit *laksha* and is the same as the Hindu word *lakh*, which means "one hundred thousand," thus graphically suggesting the countless thousands of insects which infest the trees.

The dye matter itself is derived from the gum-lac or viscous fluid, formed in punctures made by this insect, which bores into the bark of twigs and branches and becomes enclosed in the exuding juice which slowly forms a cellule, hardening into a resin which completely surrounds the insect. This substance, which has long been known as stick lac, usually attains its full size in March. The insect then has the appearance of a red, oval-shaped, polished, lifeless sac, entirely filled with a beautiful red liquid. The young lac, which grows on the body of the dead mother, temporarily uses this red liquid for food and, crawling out, leaves the body of the mother in its resinous shell. The branches of all trees, on which the lac insect reproduces its species, are almost completely covered with a brown-red crust of resin. This is the stick lac of commerce and, for centuries, has been gathered in the hilly banks of the Ganges River, in India and elsewhere, and marketed as a crude produce which, although it had not been separated from the twigs on which it was formed, yet yielded 10% red coloring matter. When the resin was beaten from the twig or limb and crushed into small pieces, it formed seed or grain lac which was converted into shellac, and the hot water, which had been used to free the seeds or grain from coloring matter, was evaporated and the residue entered commerce as a red dye, after being compressed into small,

The Mineral Dyes

The use of minerals and mineral earth substances for dyeing purposes was not commonly practised in ancient times. Quite possibly, at a very early date, primitive man discovered the art of staining fiber in local springs, found in many countries, which contained iron salts in solution for, from this source, they, quite effortlessly, obtained pleasing shades of orange and red-brown, after the liquid had been evaporated by air. Accordingly, in many different parts of the world, people learned to dip cloths in these springs, but when iron became a common metal, it was found that any soluble salt of iron could be made to act as a dyeing solution. Our colonial ancestors made these simple colors by carefully saving all scraps of used iron which they immersed in a barrel, half filled with vinegar and water, and as the iron slowly dissolved in the acid, homespun cloths of wool and linen were soaked in the solution and became yellow, orange and even brownish-red, depending upon the amount of iron absorbed in the fabric.

Green was perhaps the oldest color known to the Egyptians and, in prehistoric times, was used for painting around the eyes, a practise which doubtless had its origin in magic. This dye or pigment was powdered malachite which occurred in oxidized portions of copper lodes both in the eastern deserts of Egypt and in the hills, west of the peninsula of Sinai, in which latter place, as early as 3000 B.C., mines were worked by slaves, convicts, and prisoners of war. Blue was obtained from azurite, which also occurred in malachite deposits, but blue dyes or pastes were not used by the Egyptians until the 5th dynasty, although they are listed by Vitruvius, Theophrastus and Pliny as an ancient dye. It is highly probable that search for malachite and azurite preceded copper mining on the Sinai peninsula, for the use of both these colors preceded the Copper Age in Egypt. Palestine was without metals, for mines did not exist in the land inhabited by the Hebrews, but here and there occurred small, superficial deposits, known as pea iron or meadow iron, and these were probably used in the manner learned in Egypt, which was the source of mineral dyeing knowledge for many thousand years.

Ochres, which occurred frequently in Egypt and were among the oldest Egyptian colors, varied from pale yellow to brown-red and violet. These were obtained from native earths, colored with iron oxide. Black was always carbon, probably soot, for there is no evidence of any knowledge of charcoal. White was obtained from chalk or gypsum powder.

Bibliography

Bolton	*A Manual of Wool Dyeing*
Exmuth	*Dyes and Dyeing*
Hall	*Dyes*
Higgins	*Dyeing in Germany and America*
Hurry	*Woad Plant and Its Dyes*
Hummel	*The Dyeing of Textile Fabrics*
Keller	*The Origin of Textiles*
Lardner	*Dyes and Dyeing*
Leggett	*The Art of Weaving*
Mairet	*Vegetable Dyes*
Ostwald	*Color Science*
Salzman	*English Industries in the Middle Ages*
Textile Colorist	Various Issues since 1879

COACHWHIP PUBLICATIONS

COACHWHIPBOOKS.COM

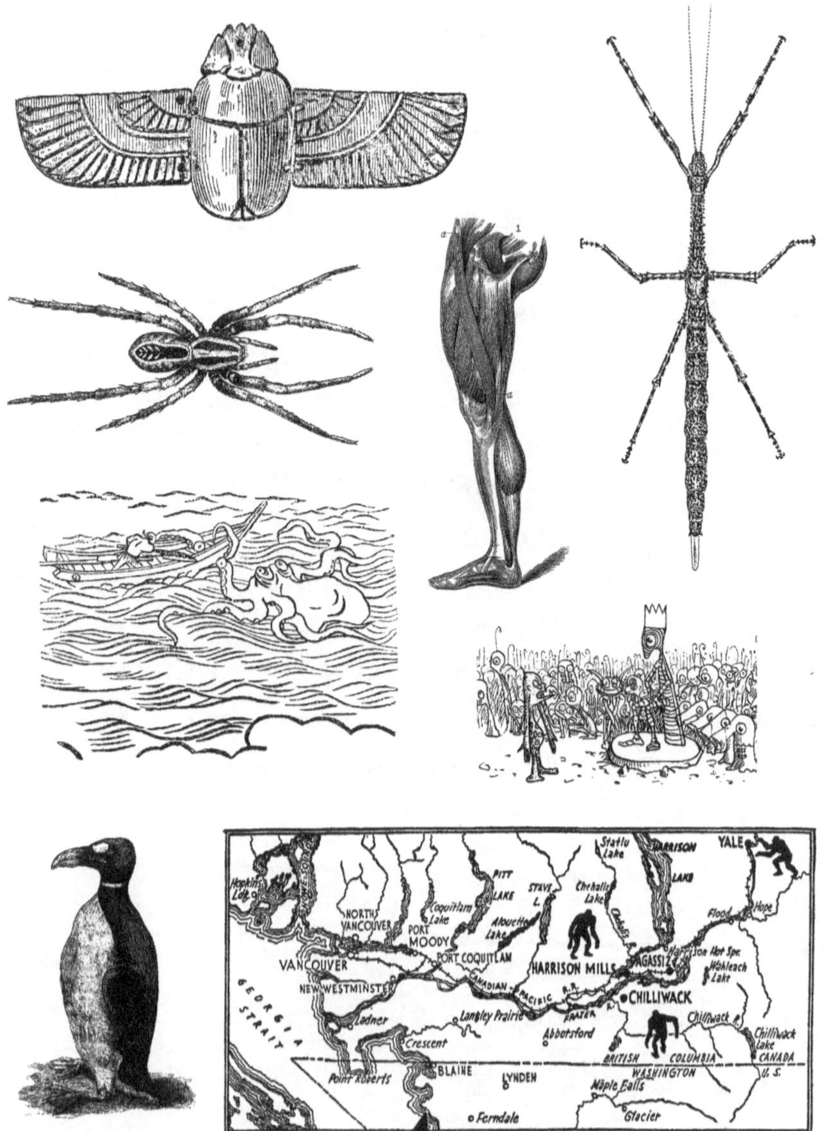

Also Available

Phantom Bouquets:
Historical and Modern Methods
for Skeletonizing Leaves

ISBN 1-930585-64-0

www.ingramcontent.com/pod-product-compliance
Lightning Source LLC
LaVergne TN
LVHW050941080826
845145LV00004B/1355

* 9 7 8 1 9 3 0 5 8 5 8 9 8 *